Daniel Garwood's

Rm. 1309

Wider Perspectives Publishing ¤ 2024 ¤ Hampton Roads, Va.

© 2024, Daniel Garwood, writes as Doowrag Daniel
1st run complete in February 2024
Wider Perspectives Publishing, Hampton Roads, Va.
ISBN 978-1-952773-81-5

Foreword

RM. 1309 is not just a book, it is a place. A door that we must all find our key to. It is a room I stumbled upon and now, I find myself going there in search of creation. A room that has produced every poem in this book and more. The air in RM. 1309 will create a chemical imbalance within you. It morphed me into a version of myself I don't recognize when I read these poems. It feels as though I wrote none of these. It feels like I was merely a brush to paint these perspectives for you. It feels amazing. How different a world looks from angles and perspectives far from what we currently allow ourselves to see. RM. 1309 has no ceiling, just constellations. No walls, just mirrors. No gravity, just peace. It does have a fridge full of food for thought, furniture made from the morals we rest on, and a pen for anyone brave enough to breathe its air and evolve into a poem. But first you must find your key. If you are struggling to find it, then consider this book a knock and a glance into RM. 1309. Nice to meet you, neighbor.

~ MR. 9031

Contents

World Builders

My momma is the only person who ever told me it was
OK to cry
Over nothing
She said cry and come see her when I'm done
I went to see my mother after the flood gates had closed
She looked at me
with eyes chemically engineered to love
She grabbed me
and held me
and told me
Never be ashamed of who I am
Never be ashamed about the size of my heart
It's a gift she said
My mother taught me
that wearing my heart on my sleeve
makes it easier to touch
others
and touching hearts is a reward worth any risk

My sister used to get in trouble for
"talking too much"
Or "being disrespectful"
People notoriously didn't like her tone
My sister is revolutionary.
My sister is an activist
My sister has always had Assata's tongue
Her voice echoes above all that dare to wrong another
Regardless of the size of the opposition

Regardless of the potential consequences
you can count on her to tell you
about yourself

My sister taught me to how to stand
She taught to have a hard head,
concrete boots and
to speak as loudly as I want
in whatever tone I want
She showed me that being a bystander is
never an option

My cousin taught me about sacrifice
She was the type of person who had quite literally
given someone the clothes off her back
Would literally trade her last
to see me smile
She taught me how to be a spoon
She said, "Love is like jelly, Daniel"...
Plenty of it for everyone
We just gotta spread it around

See, they showed me what women are.
They showed me the immeasurables
The innate abilities
That I didn't recognize until later
Like how my mother worked a job she hated
in an environment bred for negativity
But still made time to teach me
how to love
Or how my sister was the only person to ever show me
how to fight with my words instead of my fists
And because of that I haven't lost since

And my cousin had 5 kids
Instead of calling that good
she made room in her heart for me
and anybody else that needed it

See, they showed me that women are amazing
More amazing than than I could ever imagine
That women deserve praise
So fellas, protect our women
Tell your mama you love her
Listen to your sister
Look out for your cousins
Every single woman is gifted with some lesson
that a man could never teach,
but we all need to know; If men built my foundation
Then women are the caulk keeping me together
Equal in value
In importance
So never waver in our respect for women
Without the women in my life
I wouldn't be here,
But I guess that
applies to
the world
doesn't it?

Confession

Confession: I refer to my skin tone as caramel
     instead of carmel
     because it sounds more like me.
I don't really care about the proper pronunciation.
A technique passed on through my lineage,
     ain't it just like a black woman to add a little something
     something to a recipe
or a black man to skip over the instructions
     and build it his way.

Confession: I am unapologetically black.
Meaning part of the reason I chose to wear the palm tree
     I inherited on top of my head
     is because I know the shade it creates
     makes them uncomfortable.
Meaning my left knee bends to a 135 degree angle
     in time with the pendulum swing
     of my left arm when I walk.
Meaning I don't shake hands
     I dap motha fuckas up

Confession: I do see color.
     As everyone should.
     Pretending everything is black and white
     makes it easier to not appreciate the rainbow.
See Roy G Biv does not have to die for equality
     even though his rightful place is across the sky.
Color blindness causes confusion.
     Makes them confuse Malcolm, Martin or messiah
     with martyr

Confession: I hate when black comedians joke about black history.
And I know you all consider it a positive
      to laugh at pain.
But the ignorant white folks confuse you with a textbook.
Some of them think our history a knock knock joke
      with a killer punchline.
      A joke that they are dying to hear.
      A what do you call a blah blah blah Rodney King gag.
      A have you heard of the one about Emmitt Till.
      Or have you seen this Harriet Tubman meme.
The next time I say the word slave
      and I hear, "but have you seen so and so's stand up,"
      or I mention Rosa Parks and catch a motha fucka smirking
I'm swinging.

Confession: Cleopatra was buried iced the fuck out
in Cuban links and bezels on top of bezels.
      Gangsta as fuck

Confession: I say nigga too much.
      I mean I say nigga to much of y'all nigga's disapproval.
Yea I recognize it was nasty shit
      used to put us down,
      but black people been taking the nasty shit they give us
      and making it cool since forever,
      like chitterlings.

Confession: Living scares me.
      All of my idols and role models had their wings cut
      because of their rare pattern.
      Their ability to mesmerize.

I feel like a caterpillar bursting at the seems of this cocoon
    almost hoping it holds a little longer.
    I know my wings will be just like theirs
    or worse
    better.

Confession: ain't it just like a black person
    to be scared to rise

Confession: but ain't this how they wanted us to feel

5

Just because someone has the power or capability to destroy the
entire world and chooses not to
does not make them a hero.
Because you choose not to act on or say
the terrible things on your mind does not make you
a nice person.
Your evils being lesser than someone else's
is not an excuse to not even attempt to right them.
I say all this to say that
Monticello is on the back of the US Nickel.
And Thomas Jefferson is on the front.
Monticello was slave plantation that held hundreds of enslaved
Africans captive.
Thomas Jefferson built it.
Usmint.gov describes Monticello as "Jefferson's Virginia home."
When you google what is special about Monticello
it will tell you "It is an architectural masterpiece."
When you look up Thomas Jefferson
you will likely not-so-easily find
that he has black descendants as well as white ones.
If you look up why that is you will find the name
Sally Hemmings.
Sally Hemmings was a slave at
Monticello.
When searching Sally Hemmings and Thomas Jefferson
damn near every single site will speak of him
"Fathering" 6 of her children
and describe them both by using the word "relationship" or
"controversy".
Not only was Sally Hemmings a slave,

but she was also 16 when she bore her first child.
And Jefferson was 46 when he impregnated her.
Meaning there is absolutely no controversy or relationship.
That is rape.
US nickel has a kidnapping
slaveholding
rapist
on the front.
And the place that he did those things at on the back.
Thomas Jefferson is known as the third president of United States
and the guy who wrote the Declaration of Independence.
Not a rapist.
Monticello is known as architectural masterpiece.
Not a crime scene.
Makes me wonder.
What is Mt. Rushmore?

Water Cycle

How do I compare you to any one thing.
>When you have a way of becoming anything.
>one moment you are the first drop of rain to touch the hot
Sahara dessert.
>At first glance a cause for celebration.
>Then we realize you evaporated on contact.

DEATH

I remember the day everything slowed down for me.
The day I stopped talking as much.
The day I knew that learning and listening were more important
then than they would ever after be.
I was only a child and on that day
A voice came to me in the same tone as my first grade teacher
whom I thought could do no wrong
And said.
You are going to die. You are going to die for something.
So please understand that I always knew this would happen.
I just didn't know that it would be for someone.
But the more I think about it I see that you deserve it
You deserve my life.
It's crazy because I always wondered if I would be jumping in
front of a bullet or becoming the second coming of Malcolm X.
I never thought in a million years that I would be the problem.
That dying would be to save someone else from me.
I never thought that one day I would see my life as an issue.
As a burden that can only be fixed with the ultimate solution

Define ME

Pain is defined as unpleasant bodily sensation
or complex of sensations
that causes mild to severe physical discomfort
and emotional distress

Suffering is the state or experience of one that suffers
Suffer is defined as to submit to or be forced to endure
But there is no definition that describes how I felt
when you said goodbye

The closest thing is delirium tremons
Which is
a violent delirium with tremors
that is induced by excessive and
prolonged use of alcoholic liquors
And I know you're not a liquor

But that would explain how the pain reached my kidneys.
That would explain why my insides feel poisoned
Why I've become obsessed with mouthwash and hand sanitizer
and other things that remind me of you,
but aren't quite
like warm blankets
Crying
And drugs
Why?
I want to know why you left
so I can come up with ways to gaslight you
back into my arms

Even though I never truly deserved you
I promised I would crack the code on how to do that.
I was just too busy loving you
more than I love myself to get around to it.

But I guess that's what poison does to the body
Turns people into subpar versions of themselves

But I'm clean now
and
I promise I can do better

Just please
Tell me you love me again
Tell me I forgot to get your fruit from the grocery store
Tell me to grab you some wings on my way home
Tell me the door is unlocked
Tell me that guy I saw you with is nobody.
Tell me that there is still a chance
to be who I should've been in the first place

But don't tell me
You already ate
Don't tell me you live in the city now
Don't tell me you'll always see me as who I was
Don't tell me he's perfect
Don't hurt me like I hurt you
I know the brick road of broken hearts
Leads to a picket sign
That reads
"Welcome to the Land of Hopeless"

Room 1309

... And since you made it here
You can't make it anywhere
I know the trees give off pain instead of oxygen
The wells are filled with suffering instead of water

I can't go back there
So I don't care if you lie anymore
Just please... Tell me you love me

Lunch Meat

My people never make it to their death beds.
     We die in the streets.
     We die handcuffed in the backseat.
     We die screaming.
     We die walking.
     We die in ballrooms.
     We die at the station.
     We die in holding cells.
     We die in our neighborhoods.
     We die at home wondering what more we could've done
                                to get evicted.

     We die in our own beds.
     We die on TV.
     We die on Twitter.
     We die in movies.
     We die on Fox, CNN and Good Morning America.
     We die so much.   So often.
        In so many different ways.
           That at the war council
               we talk less about how to fight
                   our way out
               of the belly of the beast,
               and more on how
               to not get eaten in the first place.

Sweetest Flowers

You are proof
that the sweetest flower
does not taste how it smells.
It does not bloom
like the pictures in the catalog.

I don't know if it is chivalry
or naivety
that made me believe
that all girls are daisies
that someday become roses.
That even when you wither away
you'll leave dead pieces
of yourself and call them
rose petals.

But you were always meant to die
like flowers,
but we buy them anyways
because they have pretty smiles
and they smell so good we name
scents after them
and they look really
good in pictures
and they brighten up
your room and day and life
but then later comes,
and you realize that the scent you
loved so much was the smell of it dying.

That the pretty smile was a front for its depression.
That it used the last of its life force to brighten up your day
because it couldn't wait to die.
That it never told you it was going to die
because you never asked.
That know matter how much water and sunlight
or love
you give the sweetest flower
it doesn't matter
because
because
a plucked flower
is always doomed to die.

I was selfish
I wanted you all to myself
so bad that I was willing to kill you
for it – the flower.
I'm sorry,
even though that won't bring you back.
I'll never forget your sweetness.
I'll never forget your petals.
At least until the spring.

Hangman

Most of my friends don't like playing games with me anymore.
Not after last time they say.
Sometimes you take the fun out of it they say.
All because I said I don't like to play hangman.
Can we hang a piñata or something?
No, the game is called hang man, they say
Can we not use the color black: maybe a green or yellow?
No, that's not how we normally play
Can we just not put X's on the eyes?
No, because that's how you know he's not alive, silly
Why does this game bother you?
Because y'all see a stick figure,   And I see Nat Turner
Y'all see a children's game.   I see strange fruit
Y'all see fun,   I see one of the worst
                                            depictions of black history
                                                                  ever.

They say it's not about race
                          I say thousands of black boys were
                          lynched in the past century on trees
                                               like that one
                          I say people pick A E I O U first
                                because that was the sound
                          of pain, death, and broken necks
                                 Boys like me died that way
                                  and there's pictures of it
                                         And I've seen them
                                         and can't unsee it
I don't understand how this was made to be OK

I assume they only called it hangman

'cause lynching didn't make it past the focus groups
at the klan meetings
I hate this fucking game man
Because when you draw it out on the board
I feel like I'm in the audience
As if I'm in a scary movie I say

"Hangman is real, guys! You gotta believe me!"

They say
"Chill sometimes you gotta have fun with history"
I say, "Ain't no Holocaust game, though"
They say, "That was different"
I say, "You are conditioned to white-out America's sins
by converting it to a children's game"

They say, "Ok, then just don't play   I say "Ok,"
Sometimes I do play, though,
But my first letter is always Y

Caesar

Revolution is a word
that has garnered power the likes of no other
      within the black community.
A word seemingly conjured up through the voo doo ceremony
      where we mix brown sugar
      with Burned Sage and cocoa butter.
A word that was criminalized
      and branded
      with the same no-no tattoo   that we were.

And nothing has changed.
That word still scares people.
It is cursed with bronchitis
      so that whomever uses it
          can't help but say it with their whole chest.
So watch.

Tell your neighbor that you are joining the revolution.
Watch   as the hands of fear pull up their eyebrows
      and pull down their jaw.
Watch   as their fight or flight mindset engages
      and struggles to coordinate a cohesive sentence.

Ohhh oh ok.   I mean that's good right.
I mean I heard about that I just didn't know.
Then watch   as their ancestral instinct conveniently
      conjures up   somewhere for them to go.

They are educated.   So they know that revolution is
        defined as an activity or movement designed
        to effect fundamental changes
        in the socioeconomic situation.
See, they know that revolutions are never suggested
        they are threatened.
And ain't no revolution ever happen
        without confusion.
                Like is this sweat on my face or tears?
                Is this dirt on my hands or blood?

So we be the bad guys for now
        because history will always paint the victors
                as the righteous
        and the revolution
                as primitive.
So that no matter how right we are
        for breaking out of their cages
                standing up and saying no,

                they can call us the planet of the apes.

Room 1309

Government

Why don't you name your demons?
At least the ones that follow you.
The ones tattooed to your
Soul's report card.
At least a nickname.
Every scary movie kills the
demon By calling it

by its name.

Even demons have a
problem Calling it
what it is.

No different from us to be honest.

So name your demons.
We both know you will never get rid of them
Completely.

We both know that your demons
Are the kind that follow you to the
grave   Which

Has always been sort of their plan
So name them.
It's the safest way to live.
Call them what they are.

I roll around with
Guilt,
Shame,
And Fear.
We're all friends here.
And just like my kinfolk
They hate it when I call them by their government.

JD

JD wrote the book on disguising depression with talent.

He just didn't know he did.

JD was everything we all want to be.
The athlete that didn't go pro because he didn't feel like it.
A part of all the cool clubs
      and cool enough to make the not-cool ones cool.
JD was not the jack of all trades,
      but more like the prince
      who we couldn't wait to see become king.

But JD hated being talented.
He hated being a genius, because every time he showed how
special he was it reminded him that no amount of love, fame or
attention could fill the hole in his heart.
Who knew,   the most amazing people sprout up
      from the quicksand of desperation.
The reason JD was so great at hockey is because
      he thought it would fill the void,   but it didn't.
As with the drama club,
      the school paper,
        his grades,
          Stacy and Monica and ShaiAnne

      nothing he throws his heart at fit that crevice in his aorta.

Until his fingers found the strings of the guitar.
JD entered a different atmosphere when he played.

Time stopped the moment the smooth oak of his acoustic
        touched his thigh in a way that no girl ever could.
When the sling of his guitar strap wrapped him up
        in an embrace tighter
        than his his family could ever hold him.
The strings.
The strings
The strings worked together to guide his fingers
        to a hymn of a lost part of himself.

        The guitar filled that hole in his heart.

But it was too late.
He had already beaten his heart
        loveless to a point of no return
                by recklessly giving it to things
                        that would never appreciate it.
Things that could never love him back
        like hockey,
                the drama club
                        his grades,
                        Stacy, Monica, Shai Anne.
Now we catch JD punching holes in plaster after he writes a song.
We find him crying in the fetal position
        5 min after cracking DaVinci's code.
JD never learned to protect his heart from parasites.
He sent his heart off to war on a journey to fill the hole
        only to return and find the termites have caused more.

        But JD is still great at everything.
        He just plays sad songs now.

Room 1309

Venus

You are a different kind of flower. A Venus fly trap
that snaps shut with understandably out of this world
speed. You are surely a long way from home.
A couple thousand thousand years away. Not quite a light
year but bright enough to sink my attention deep
into the thorns you call lips. And slowly, digest my heart.

My mom told me that my sleeves were no place for my heart
and yet here I am. 23 years old and still trapped
in the ways of my teenage self. Love is staring at the deep
end of a pool from above. Jumping is exiting one world
and entering another. Swimming is finally seeing the light
in your eyes. I can't swim but I never leave my sleeve at home.

The tips of my fingers have become its new home,
now that I've realized my mom was wrong. I keep my heart
out here so when I touch pretty things that aren't quite light,
they feel all my love. You are a snare trap
and my instincts tell me to jump through hoops for you. A world
where my heart is my skin means the slightest cut is too deep.

Which is why when your thorns puncture my love it's so deep,
that I almost regret wearing something so fragile. "Home
is the safe haven" mom said. But your magnetism makes my world
rotate on a different axis. One that makes my reckless heart
feel acceptable and makes your beauty feel less like a trap,
And more like the space at the end of a tunnel that makes light

I just don't know if I should run to or away from the light.
I'm leaning heavy towards to though. You are Mariana Trench deep
in the busy places of my mind. I'm not mad I fell for this trap
because beauty is the perfect bait. I'm scared at how easy home
became synonymous with your presence. How quick I gave my heart
to something potentially venomous. How you devoured my world.

How much I love all of it. How much I truly deify this world
you amassed by absorbing and taking into yourself the light
of my soul and combining it with the purity of your heart.
The sudden shift of life for me is scary but I'm in too deep
to even want to escape your inescapable grip of love. I'm home
now if you ask me. I love the way I love even caught in a trap.

This flower from Venus has shown me how deep in beauty my home
can be when I let the fear in my heart blossom into light.
This trap isn't the end of me but the beginning of a new world.

Dear Black Woman,
(the response)

I love you too
In all your shapeshifting
In all your essential oils and shea butters
In all your sage burning and incense lighting
Black woman
I know god gave you them hips to prepare you for all the
burdens this world has waiting for you.
I know you capable of carrying all that
and then some
But please
Let me get that for you
Allow me to hold
that weight So that
you
May stride like the queen you were always meant to be Yes
Black woman
You are a queen
Not some princess trapped in castle who just so happened
to marry a prince
Nah
A queen
Meaning you run shit and you walk like it
The crown matches your red bottoms and you talk like it
The world is ignorant to your royalty
Black woman
In a world where the wage
gaps Are proportionate to the
thigh gaps
You runway strut into CEO positions anyways

In a world that would prefer the comfortablity of the
silent woman
You
Speak loudly and carry the big
stick You do it all black woman
The first barbers for the black sons
Carpenters and plumbers when you gotta be
party planners he'll bent on making the babies smile
Scientist botanist nutritionist Therapist mathematicians
And beauticians
Are some of the things on the never ending list of things you
make look easy on your spare time

And sistas
On behalf of all the black men
Whose tough love upbringing
Cultivated our hard heads
I'm sorry
For the times we've made you feel like less than a queen
For the long days and nights that you felt alone
Your are not alone sista
When you trip stumble or fall it'll be a black man like me
standing between you and the pavement
Trust that we will treat any jewels you drop like fine china and
devote our livelihood to catching them
And I promise you black
woman That I am not
alone.
That I am not a rare breed

There are brothas like everywhere who are devoted to the
cause of uplifting our sistas in need.
And we understand there are those who are lost
Those who struggle to comprehend your identity
Those who grew up with hoop dreams but never learned
how to court you
We will teach them
Black woman
Teach them how to love you
How to listen
Why we cherish you
Why we protect you
Show them that we could never be
Without you and all your glory
And all your sacrifice
Black women stand tall and never look back
Cause we got that for you
Fuck what they say about your hair
Because we love that on you
When you tell us that you love us
Trust we'll love back on you
Tell us your your favorite dish
And then we'll cook that for you
Ask me something I don't know
Imma learn that for you
And if you find yourself in a situation
Where your struggling with something
That is not worthy of your stress
Call me
And I'm a' get that for you

Dear City Girl,

Why do they call you city girl?

Is it because you are naturally drawn to what's known as
outside?
Is it because other
peoples extra Is your just
enough?
Is it because chaos calms you?
You don't look for it
But you always find yourself at the center
Among the skyscrapers,
Street lights, and open signs
In the center,
Surrounded,
Where anyone from white collars To dog collars can get along.
Where the music is always extra
loud Just loud enough
Is it because you love to dance?
If Bruno Mars taught us anything it's to let city girls be

city girls

Dear City Girl,

I bet your scars have better
stories Than your tattoos
I bet you're more than just a city girl
That's probably just a piece of one of the counties that make
up one of your states of being
I bet your more like a continent
Or body of water

Room 1309

Dear City Girl,

I read in a book somewhere
That city girls don't chase they get chased
And I think I understand why
I understand how the bright lights can
Masquerade themselves as the carrot at the end of this
hamster wheel
And how I may very well
be just another pet who dies on the wheel
never quite getting but always left wanting

Dear City Girl

Any man would be lucky to see you undressed

To grip those curves
And kiss those lips
But I
Wanna watch you meditate
That's how I really see you naked
Grip your attention
Pucker my lips to French kiss
Your inner city girl

Dear City Girl,

I don't know if you'll ever get this message

But if you do
Write me back.

Mr. Softie

When the ice cream truck of life stops at your home.
And you show up to pick your flavor.
I hope you pick soft serve on a waffle cone
With cookie dough and caramel drizzle
Instead of the prepackaged
Pre-made Spongebob or Sonic The Hedgehog
With gumdrops
'Cause frankly
I'm nothing like them regular niggas, I'm made fresh
And I always hit the spot

Panorama

You look like some perfectly laid baby hairs
You smell like lavender candles and scented lotion from bath
and body works
You sound like Giveon and Erykah Badu harmonizing on a hook
You taste like a pina colada and papaya flavored sno-cone with
pop rocks on the beach in 97 degree weather
You feel like if Tempur Pedic made hoodies and blankets
You walk like you taught Jesus how to walk on water
You move with no remorse for the imitators
Constant pressure on the necks of the casuals
You are a metaphor who should never be confused with a simile.
Sorry for my ignorance

Waterfall

It seems I always find myself at the mercy of the waterfall.
Wondering who the fuck told gravity to be so narcissistic
To be so demanding
So stubborn
A king pin in this world of life
No matter how many pills I take
I will have to answer to him
The higher I decide to elevate
The longer the free fall
The harder the surface
All because gravity has always been A hater
A Jealousy driven entity
That can't stand to look up.
Can stand to see me climb.
Has to find a way to bring me back down to its level. But I
Have always had a thing for verticality.
Never been one for the long jump
Always one for the pole vault.
But gravity has never lost
Falling is inevitable
So I must fall with grace
I must hypnotize with the beauty of how I Hit the surface with
everything I got.
No skydive. No parachute. And ain't nothing free about it.
Just waterfall.

The Girl From Chula Vista
& Her Jansport.

The love connection that we all wish we had.
Trust.
Like trust trust.
Like the only thing in her life that she ever trusted to truly have
her back.
Like The Skyrim ally that seems to be able to hold everything
she doesn't feel like carrying like the weight of her anxiety.
she bottles it up
but whenever she tries to hand it off
to one of those "oh you can talk to me about anything's"
the weight turns red
and they are somehow incapable of holding it,
but not her Jansport.
Jansport doesn't ask where they're going
or why she's crying.
Jansport won't try to one up her shitty day with
a dramatized story of its own.
no.
Jansport just opens its arms
holds on tight
and has her back.
The girl from Chula Vista always knew a jansport was
what she needed,
but could not afford one so she settled
for whatever fit the budget the same way
she chooses her relationships.
But when she gained the money she understood
that choosing a ride or die

is more than a purchase, it is an investment.
An investment that will have to be reliable enough
to not allow her depression to hit the ground.
Reliable enough to not forget her smile at home.
To always have her joy within arms reach.
To carry her deepest secrets.
To conceal her wallet, tampons, and butterfly knife
and never tell a soul.
Jansport has traveled the world with the girl from Chula Vista
and never left her back
not because she has no friends.
But as an example of what a friend should be.
She does not want a BFF right hand
or self proclaimed sibling.
Just a Jansport.

Dirty White Converse

To the guy chasing the girl in the dirty white converse
Stop

Let her walk away

There are so many red flags associated
With the dirty white converse like

A girl's shoes are a good reflection of a girls room
And a girl's room
Is a direct reflection of a girl
And dirty shoes don't live in a clean room
It doesn't match the aesthetics
And dirty rooms are a sign of immaturity
And chaos
And lack of responsibility
Meaning she's probably a jit
It's also a sign of misdirected loyalty
Which is subconsciously what attracts us to them
We love the fact that she's capable of being a ride or die
Literally rolling till the wheels fall off
But best believe
It's laced in toxicity
Because

She won't want you to improve or clean yourself up.
She'll make shit up and gaslight
By making you believe she loves you just the way you are. She
won't take responsibility for dragging you through the mud
When all you wanted to do is support
her Sole

Daniel Garwood   38

The one thing she uses you for.
Chucks are easy to keep clean.
All you gotta do is give a fuck
So to the guy chasing the girl in the dirty white converse

And the guy chasing the girl in the dirty white forces
And the guy chasing the girl in the dirty white 1s
I hope you're willing to go through all that those shoes have.
If not
Keep walking.

Apnea

When darkness calls
It's voice is the offspring
Of a banshee and a siren.
Impossible to ignore.
Loud enough to penetrate the mind
And psychologically convince your body
That it is heavier.
That one more step is one too many.
That your next blink could and should be your last.
Darkness has a very demanding tone.
A very seductive resolve.
It convinces you that it knows best
That rest Is exactly what you need.
It reminds us of the abyss it provides.
That space where every problem
Every stress
Everyone
Has been Thanos snapped out of existence.
All that exists in the darkness
Is your wildest thoughts
Your moonless subconscious perspectives
And it is temporary.
Sleep can be chamomile
Or a tea cup's sunken place
It all depends on your sacrifice
To the dark side
Because nightmares are dreams too.

Insomnia

One, two, Freddy's coming for you
Three, four, better lock your door
Five, six, grab a crucifix
Seven, eight, better stay awake.
Stay awake I tell myself.
        I have seen what happens to those who sleep.

M.U.

To those that don't quite understand
when I say I'm medium ugly,
I do in fact mean it with confidence.
I am a proud member of the medium ugly community.
Medium ugly meaning my natural beauty lies on the inside.
I know ever since I was a kid
I've been that one short light skin dude.
Nah the one with the pretty eyes.
Yeah,
him.
And it took a roller coaster of toxic masculinity highs
and utter insecurity lows to crack the code and finally understand
what it means to be medium ugly.
Medium ugly is not complete confidence or complete
apprehension it is a state of mind
that is scientifically known
as no fucks given.
Meaning, I reached a point
where I don't see competition within myself or others.
I don't desire to be sexy or good looking anymore
I desire to be me.
Under 6 foot with a metabolism
that won't allow me to weigh more than 160lbs
Medium ugly means smiling
because you can't find a reason not to.
I am not Michael B Jordan.
But I am Daniel L Garwood though.
And I'll never be anything else.
Medium ugly is comfortable as fuck.

Let Me Count the Ways

How do I Love thee?

       Let me count the ways

See, if Cupid's Arrow pierced my heart and you were the
first thing I saw,
Nothing would change
Because loving you is business as usual
Because your arrow was venomously concealed
And sealed with a kiss all those years ago

       But let me count the ways

I love you selfishly
I love to see you laugh unless its at someone else's joke
I melt when you smile unless its at someone who isn't me

       But just let me count the ways

I love you relentlessly
We went from across the city
To across the state
To across the country
To across the world
Where I told you I loved you and that you were the most
beautiful girl in the world
At every stage
Through every hiccup
Through every struggle
Through all the pain
Through night and day
Through all the time
Through all the distance

       I loved you and won't stop

Room 1309

But please just let me count the ways
My love for you is a gift too big to fit in your car
So I'm afraid you won't take it
Or even worse
You'll accept it
Realize it won't
fit into your life I
mean car
And eventually give it back
But there's no warranty on love

It's a no-down-payment
As-is type of thing

And I know I know I know
But please just let me count the
ways I love you figuratively
Literally

Passionately
Haphazardly
Practically
Casually
Erratically
Emotionally
Lustfully
Devotedly
I love you with intensified intent
Fiercely
Profoundly

I know I'm kind've rambling, but I got more I promise
So please, just please

Just Let me count the ways
I know, I know
I can make you understand
I know the word or phrase that perfectly describes it is in
here somewhere
Just please wait
Wait just a moment longer for me to find it
Give me a second to search for it
Just please
Let me count the ways

The Matrix

What if I didn't pick a pill?
I was force fed both
And now I'm trapped in this…
Inter-societal purgatory
Where we know what the world really is
And yet we still pretend everything is alright,

But it's not

They still look at as
The monkeys in the people clothes
The affirmative action life acceptees
We don't belong here.
No matter the politics
No matter how many steps our Fitbits say we've marched
We will not claim this territory.

For we are not welcome
And it is shown in their idols.
Like how many slave holders are on US currency?
How many black descendants
have an ancestry linked to Mt Rushmore
Because of the constant rape and sexual assault
of African women?
How do we now chastise the system of slavery,
but not the men who built it.

How much of this is public record?
So long as history writes the evil-doers as heroes
We will always be the villains
Or at most the vigilantes

The matrix is such a weird place
Especially when we know we are in it

But we don't know what our other option is

Solstice

Progress
Is just a prepubescent sunflower
In the heat of summer
Promising

Brightness and positivity.
At least until the solstice.
When all the living things stop growing.
When the pretty petals fall to the ground
Will you still see beauty?
When the bright green stem fades to a decrepit gray
Will you still see positivity?
We should
Because struggle is the prettiest phase of progress
It takes you to the bottom of the pool

So you can understand the possibility of drowning, but also

So you can catapult yourself to the surface
There's beauty in those gasps for air
There's beauty in the journey to the surface
There's beauty in the sunflower
No matter the season
No matter the solstice
Because progress is not how hard you fall Or how fast you get up
It's the distance from the bottom to the top
From the roots to the buds
From where you came from to where you going
Hell to high water
Death to beauty
Solstice to solstice

Eclipse

I will never say that you are self-centered
Just because the world revolves around you.
Because honestly
I feel like that's what worlds are supposed to do.
I imagine this is some sort of ceremony
where we circle, chant and worship that which
is Absolute.
That which the universe is built around.
Sometimes all I can do is
Hope that the axis I spin at
Allows enough time for your beauty
Radiate both of my hemispheres and
Power my mind enough that when
I inevitably see your dark side
There is enough of you stored to sustain life.
Until the eclipse is over.

Guilty

The toughest revelation that I have had to come to terms with is
that I,
alone,
can not help everybody.

There will be times when I am not enough.
When the person I am will not just be of no use,
but I will be volatile to the situation.
I am not superman.

Not because I wasn't born on Krypton
or because I don't have a cape
or an alter ego
or super powers but because the man himself.
The one that seems to do no wrong.
The one that finds a way to be of use in every scenario.

The hero
the vibe
the charm
is fictional.
It's all meant to be fake.
Yet
as I understand this
It still feels like brushing my hair backwards.
Swimming against the tides.

It is going against the grain for me to be anything other than real.
But as I get older I realize

That no matter how pure my intentions are there will be moments
when the person I am is the sharpest blade in the room.
The biggest gun.
A weapon to some unexpecting soul.
And I will be the one doing the damage.
Doing the hurting.
Causing the pain.

My hugs will be guillotines,
Kisses filled with venom,
Words breathed out of my mouth in a cloud of toxic gas
And then the one with the purest heart becomes the serpent.
Hercules becomes Hades.

I will have to answer the toughest question about myself.
Does the pureness of my heart
And the love that I have outweigh the damage done,
Or the people I have hurt.
Do my intentions even matter if the outcome is pain.
Am I a good person, or a bad person?
And while I put my soul on trial
I must figure out who plays judge jury and executioner.
What evidence will the prosecution present?
Who will the prosecution bring to the stand?
Who is the prosecution?
But as tough as it is I understand the truth
I know the outcome.
Guilty.

Palm Trees

In 2019, I deployed out of Norfolk into the Middle East.
We took the Atlantic Ocean to the Suez Canal.
It was the first time I saw Africa.
The first thing that came out of my mouth was, "Damn, I ain't
know we had Palm Trees."

Growing up in Detroit, Palm trees always just seemed so exotic
and expensive, and all the things America teaches you Africa is
not.

My Black Pride morphed itself into a smile as the motherland
looked at me and said,
"Don't let nobody tell you Momma ain't still got it."

But it was really the journey that found a way to tattoo itself into
my memory.

As we sat in the middle of the Atlantic enclosed by the baby blue
glass, we stared out at the horizon and talked.

When someone asked what we thought could be at the bottom of
the ocean, the roundtable responses consisted of diamonds, drugs
and money.
But with a blank stare I couldn't help but say Black Bodies.
Because this is where the brave and able-bodied jumped.
This is where the enslavers decided the rations wouldn't last if
they didn't cut weight.
This is where decisions were made.
Where epiphanies were had.
Where the future was seen.

Where mother's had to decide for their own children, That death meant liberty.

I realized that somehow I was on the middle passage going backwards.

And that my urge to jump was just as hereditary as my instinct to stay put.

That even though the ocean is supposed to be a reflection of the sky the hue is darker.
Because it has been Kool Aid mixed with blood and dark skin.
I realized that I'm here.
Where they were.
Where stories of black fathers running to the store real quick originated.
Where black hope originated.
There ain't nothing like that last I'll be right back,
Or I'm going to get help.
I'll save us.
I imagine Mothers had to figure out a way to console their children.
To let them know Daddy's going to be alright.

I imagine they had a similar discussion about what's at the bottom of the ocean.
I bet they said Palm Trees

Dragon Breath

I find myself as the epicenter
For all my friend's negativity.
They know I am the best at absorbing their
Problems.

But they don't know that
That their negativity
Gets regurgitated as a fiery vomit
In the form of
Pent up anxiety and pain.

That I am a great listener

But sometimes

I find myself alone
Wondering why I am crying.
That my hugs are a cure to almost Any heart break

But sometimes I wake up

And my chest feels like it has been Hollowed out
So that Picasso could
Tattoo his love interest
On my heart.

That my voice and use of words
Had the ability to soothe, calm
And talk people out of
Choosing to walk the plank

And I paint life as
This thing that inevitably
Goes around
And comes around

But,

The nights I can't sleep,
Their anxiety brings facts, statements
And eye witness accounts

That make it hard to argue against
Making bagels in the bathtub.
See they don't know that

I am a dragon.

Cursed to a life of always
Being the biggest target.

Dammed with the ability to absorb damage.
Blessed with lungs to

Breathe fire.

Wet Vines

Life is a rainforest
And I am the king of the jungle.
Royalty out of its element.
It's easy to see the similarities

But I'm not used to wet vines.

I'm used to Tarzan swinging
From one to the next
Always knowing my next catch
Is guaranteed.

Expelling from my memory
Yesterday's vine
And going all out for tomorrow's
I'm used to being awarded
For my vulnerability.
For allowing myself
The century long millisecond
Where I am fully committed

And yet

Nothing in my grasp.
A king as committed as me
Does not belong in the rainforest.

The first wet vine
Will send me skydiving
Into a narrator's "Here lies..."
But once you take
Your first hit of flotation
And understand the constant state of high
You dread the ground.

Your knees aren't strong enough.

I hate wet vines

Room 1309

Dreamer 101

What happens to a dream differed
Or a dream hated on
Shot down with 50 caliber rounds made of
Preconceived concepts
Of impossibilities

It dies with an asterisk
That exempts my own dreams from that prophecy
Because I am the prophet
I have a prewritten conception adventure and demise
    So
Your own beliefs don't apply to me

    I was born a dreamer

    Wide asleep
    And dead awake

Dreaming is my purpose but it is far from my prophecy
I was bred
to not only dream the fairytales
Dreams that you so graciously place
On your assassination hit list
But to rewrite them as fact
To prove that we have become sheep
Content with our picket fence world
Wrongly believing
That anything outside of our realm of sight
    is merely speculation

That anything over the horizon is merely an illusion

We have broken the mold of what is considered possible
generations after generations and we will not stop now
        I will not stop now
The crazier people tell me I am The more I smile
The more I get told of naivete and innocence and idealism
I know those are all just synonyms to dreamer
 And I know I'm doing the right thing

So, to the dark clouds that walk among me as
        acquaintances,
                friends and
                    family;

Whenever you decide to curl your lips crook your neck
        and shake your head at one of my dreams
And fix your tongue to question my sanity
And place a ceiling on my capabilities
        based on your insecurities
                based on your negativity

I ask that you hold that thought   Sit back and watch
                I'm gon' show you how great I am
                I'm gon' show you how to dream

Perry

Isaiah Thomas,
Derrick Rose,
Mugsy Bogues and
Chris Paul
Are athletes who's skill
Don't quite match their stature.
Players who were always regarded as
Playing bigger than themselves.
As if their success was a puzzle piece
That didn't belong in this picture.

But Phineas & Ferb taught us
That it's OK to be a platypus.
That we are less cookie cutter
and more hand made.

That there are no such things
As imperfections

Only a bakers uniqueness.

We don't have to see the vision.
We don't have to understand the purpose
Or the method behind the creative process.
But we do have to respect its
Unforgettable nature.

Stop confusing the equation
That translates ignorance
To fear and jealousy,
With the one that equates
the misunderstood,
And The unrecognizable,
With admiration.
Because deep down,
We all know platypuses
Are kinda dope.

Faded Scents

I knew the end was near
When missing you started to hurt.
When I started having a hard time sleeping
And an even harder time walking up.

Who knew

That your peace could be so loud.
that your serenity
Could be everything
Even when all you need is

Nothing at all.

That the moment you kissed me goodnight
Would be the moment
Love equaled pain.

Who knew the good times
Were just the catalyst for the hard ones.

That blown kisses
Are just airborne venom

And now

I sit here.
With your love hovering around
The campground set up in my subconscious.
Waiting for it's SOS message to be answered
So that it can be brought back to civilization

                        But I know the truth.
That your subtle hint of lavender
Faded away months ago.
That time away is for the stable-hearted.

Graphite

I've collected all seven dragon balls,
a magic oil lamp,
and a birthday cake
at 11:11 PM
as the night sky is pierced by a shooting star to say,

I wish I was your notebook.

Not because I want you to stab me with graphite
But because I want your attention
I want more time connected to your gaze between blinks.

Your attention has been added to my list of things
that aren't sex
but feel like sex.
It's a Q-tip hitting that spot.
A medium rare steak with no sauce seared on a cast iron skillet
It's, have you ever smelled pineapple juice,
If not trust me it's that
Your attention is a shareable bag of white chocolate M&Ms to me
Hard as fuck to get for some reason but when I get it
I literally can't help but to dive right in and only when I feel
like the bag is almost empty do I remember how rare it is.
Your gaze is a gust of wind when its 98 degrees outside,
but the weatherman said it feels like 105.
Meaning its one of the few times enough is not enough
Your focus is not having to wake up in the morning
Warming comfortable and freeing.
It makes me feel like nothing else matters

Your attention is the Penelope that
Suitors from across the land desire
From as far as the eye can see
But on this odyssey
I am Odysseus

Meaning don't nobody want it like I want it,
So yeah

I've summoned the genie from Alladin,
Shenron,
some red velvet cake,
and a flaming chunk of an interplanetary comet to say

I wish I was your notebook.

Not because I want you to stab me with graphite.
But if that's what it takes.
Unless of course you're writing in pen.

Narcs

It's crazy how my narcissism shows in the weirdest times
Like how we could both be falling in love But My

Selfish subconscious competitiveness
Won't allow you to hit the ground before me
So I love harder
Only a true narcissist thinks himself capable
Of both defying and magnifying gravity at the same time
But to believe you are gravity itself
Is a different level of vein
To think I have the power to attract or repel any and everything
in my orbit
Is to think that everything revolves around me because
I allow it
That physics is not the study of energy and matter
But of he with the most energy and that that matters the most

Terriers

Have you heard the one about the dogs?

What is the difference between a golden retriever and a
Rottweiler?

Well one's a house dog and the other is not

One drinks from the tap

The other has learned to catch the droplets from the water spout.
Learned not to drink from the runoff of the gutter
Considers rain showers a holiday.
The house dog knows no chains

The outside dog is an expert on the 12 square feet surrounding
the tree

The house dog knows collars and subtle tugs at misdirection

The outside dogs knows noose knots chains
which served to punish their curiosity
and cap their freedom.

The house dog misperceives being paraded from
being celebrated
when confronted by company.
The outside dog knows exactly who it is.

The house dog knows food in plastic bowls.

The outside dog is shielded from the world by tall fences
permanently branded by beware of dog signs.

The outside dogs know the jingle bells of metal bowls
The house dog sits when told and sleeps on soft rugs
The outside dog sits and sleeps on the dirt
Historically

I wonder when niggas started calling each other dogs

Room 1309

Colophon

Wider Perspectives Publishing regrets to have to announce that the ongoing Colophon page, used to tout artists published in books from WPP, has to be reworked. This is due to the growing library of fine writers coming out of, or even into, the Hampton Roads area of Virginia. Within days of the release of Daniel Garwood's Room 1309 WPP will have published 100 books in Novel Fiction, Memoir, Inspiration, Short Story Collections and Poetry Poetry Poetry Collections.   Yeah, it's getting hard, *sigh*.

Tabetha Moon House
Travis Hailes- Virgo, thePoet
Nick Marickovich
Grey Hues
Rivers Raye
Madeline Garcia
Chichi Iwuorie
Symay Rhodes
Tanya Cunningham-Jones
          (Scientific Eve)
Terra Leigh
Raymond M. Simmons
Samantha Borders-Shoemaker
Taz Weysweete'
Jade Leonard
Darean Polk
Bobby K.
          (The Poor Man's Poet)
J. Scott Wilson (TEECH!)
Charles Wilson
Gloria Darlene Mann
Neil Spirtas
Jorge Mendez & JT Williams
Sarah Eileen Williams
Stephanie Diana (Noftz)
Shanya – Lady S.
Jason Brown (Drk Mtr)
Ken Sutton
Kailyn Rae Sasso
Crickyt J. Expression

Faith Griffin
Se'Mon-Michelle Rosser
Lisa M. Kendrick
Cassandra IsFree
Nich (Nicholis Williams)
Samantha Geovjian Clarke
Natalie Morison-Uzzle
Gus Woodward II
Patsy Bickerstaff
Edith Blake
Jack Cassada
Dezz
M. Antoinette Adams
Catherine TL Hodges
Kent Knowlton
Linda Spence-Howard
Maria April C.
Tony Broadway
Zach Crowe
Mark Willoughby
Martina Champion
... and others to come soon.

the Hampton Roads
    Artistic Collective (757
    Perspectives) &
The Poet's Domain
are all WPP literary journals in cooperation with Scientific Eve or Live Wire Press

Check for those artists on FaceBook, Instagram, the Virginia Poetry Online channel on YouTube, and other social media.

www.ingramcontent.com/pod-product-compliance
Lightning Source LLC
Chambersburg PA
CBHW071240090426
42736CB00014B/3160